Edwin C. Powell, William McMillan

Street and Shade Trees

Practical Essays on the Subject of Trees of Street and Lawn Planting

Edwin C. Powell, William McMillan

Street and Shade Trees
Practical Essays on the Subject of Trees of Street and Lawn Planting

ISBN/EAN: 9783337813338

Printed in Europe, USA, Canada, Australia, Japan

Cover: Foto ©berggeist007 / pixelio.de

More available books at **www.hansebooks.com**

STREET AND SHADE TREES.

By

E. C. POWELL,

Assistant Editor of American Gardening,

AND

WILLIAM McMILLAN,

Superintendent of Buffalo Parks.

The Use of Shade Trees.

From Nursery to Permanent Location.

What Trees to Plant.

Hardy Trees for Particular Purposes.

Shade Trees in City Streets.

PUBLISHED BY

THE RURAL PUBLISHING COMPANY

TIMES BUILDING, NEW YORK.

STREET AND SHADE TREES.

CONTENTS.

OTHER ISSUES OF THE RURAL LIBRARY SERIES.

ISSUED MONTHLY, $3 A YEAR.

THE TUBEROUS BEGONIA : Culture and Management. By numero
Practical Growers. Illustrated. 20 cents.

RATS AND OTHER VERMIN ; How to Rid Buildings and Farm of. 20 :

CHEMICALS AND CLOVER : On the Use of Fertilizers with Sod for Su;
plying a Cheaper Manure than by Keeping Live Stock. Over 100,c
sold. 20 cents.

HOW TO PLANT A PLACE : On the Ornamental Planting of Small
Places; A Brief Popular Guide. 10th edition. By Elias A. Lon
author of Ornamental Gardening for America. 20 cents.

THE BUSINESS HEN : Breeding and Feeding Poultry for Profit. By 2
Practical Poultrymen. Edited by H. W. Collingwood, of *The Rur*
New-Yorker. Double Number. 40 cents ; in cloth, 75 cents.

CROSS-BREEDING AND HYBRIDIZING OF PLANTS. By L. H. BAILE\
40 cents.

THE NEW CELERY CULTURE : No Banking-up Required. The Practice of Practical Men. By Robert Niven and Others. 20 cents.

CANNING AND PRESERVING FRUITS AND VEGETABLES, and Pi
paring Fruit-pastes and Syrups. Experience of many Practical Work
ers. By Ermentine Young. 20 cents.

ENSILAGE AND THE SILO : All About Preserved Fodder; Conserv
Cattle Food. Experience of Fifty Ensilage Farmers. Edited
H. W. Collingwood, of *The Rural New-Yorker*. 20 cents.

ACCIDENTS AND EMERGENCIES : What to Do in ; Home Treati-<
of : What to Do Till the Doctor Comes. By G. G. Groff, M. D. 20 cen

MILK : MAKING AND MARKETING. Being Illustrated Accounts
Successful Milk Farming. 20 cents.

FRUIT PACKAGES. The current styles of Baskets, Boxes, Crates anu
Barrels used in Marketing Fruits in all parts of the country. Ey
E. C. Powell, Assistant Editor of *American Gardening*. Fully
Illustrated. 20 cents.

MY HANDKERCHIEF GARDEN. Size 25x60 Feet. Results : A Garden
Fresh Vegetables, Exercise, Health, and $20.49. By Charles Barnard.
Illustrated. 20 cents.

STREET AND SHADE TREES.

PRACTICAL ESSAYS ON THE SUBJECT OF TREES
FOR STREET AND LAWN PLANTING,
WITH DIRECTIONS FOR TRANS-
PLANTING, AND A LIST OF
DESIRABLE TREES.

BY

EDWIN C. POWELL,

Assistant Editor of American Gardening,

AND

WILLIAM McMILLAN,

Superintendent of Buffalo Parks.

NEW YORK:
THE RURAL PUBLISHING COMPANY.
1893.

CHAPTER I.

THE USE OF SHADE TREES.

TREES, other than fruit trees, are planted mainly for two purposes, ornament and shade. For ornament alone we desire trees that are beautiful in color or shape of leaf, color of bark, habit of growth, character of flowers or oddity of habit. Cut-leaved trees and those of a graceful, weeping habit do not cast a great amount of shade, nor attain great size; neither are they able to withstand neglect or abuse. They may be said to belong to a higher order than other trees, and with their higher structure comes a greater and more complex development of parts, which necessarily renders them more delicate and susceptable to injury, climatic conditions and changes. A purple beech or cut-leaved birch would be as much out of place, even if it could be made to grow, in a crowded city street, as would a mammoth oak in the back yard of a 25 x 80 foot city lot. For shade purposes, then, it is desirable to secure trees which present characteristics somewhat different from purely ornamental trees. Some of them have directly opposite characteristics, others similar ones; as the character of the one class approaches that of the other the trees may be used for the one purpose or the other. Shade trees may be used for ornamental purposes, but the purely ornamental trees, so called, are not generally adapted for shade or street planting.

The chief requisites of a shade tree are that it be large and shapely, with abundant foliage, so that the sun does not shine through to any extent. A street tree must possess, in addition to the above qualities, a disposition to transplant easily when of good size, ability to grow well in poor, dry, hard soil, be capable of withstanding cold, heat and dust, and have few or no enemies. Along a country road or wide village street the soil is usually better than in a city street, where either the good surface soil has been removed in grading, or sand or other

equally poor soil has been carted in on top. Paved streets have
gutters that carry off the water, and the soil beneath is usually very
dry and hard. In the city there is also a great amount of dust,
smoke and soot, which is fatal to many trees, especially to evergreens.
The conditions which a tree meets in the city street are directly opposite to those of its natural habitat; therefore, it is not strange that
we see few large, healthy trees in the thickly settled streets of any
city of considerable size or age. Most of the large trees we do see
were planted when the city was a mere village, or they came up nat-
urally before the street was laid out. The roots have gone far and
deep in search of food and moisture, and became established before
the present conditions existed. Along country roads, in villages and
the suburbs of cities, where the streets are not paved, more of the
natural conditions are present, and dust is the only serious enemy
present. Thus a great variety of trees can be used in such places,
and better and more perfect specimens can be grown. On the lawn,
in parks or private grounds, the chief natural condition which does
not exist is the shade and protection of other trees that is found in
the forest. By planting in groups this can be partly afforded, but
most trees thrive by themselves after a year or two of slight protection. A nursery-grown tree is hardier and will stand transplanting
to a position by itself better than a tree taken from the shade of its
native forest. Never buy trees that are brought into the villages in
wagon loads from the woods and peddled about the streets ; they are
usually poor investments.

For shade purposes and nobleness of stature there is no tree in the
north which equals or excels the American elm. Its great hight
and spreading, drooping branches place it above all rivals, but the
elm-tree borer has become so serious in many sections in the east
as to almost exterminate it and forbid setting any more. The canker
worm and several caterpillars have also preyed on it, and it is such
a high tree as to be out of reach of most spray pumps. The elms
of Northampton and of Old Hadley, Mass., are famous for their
stateliness and grandeur, and one who has seen them cannot but be
impressed with their beauty and desirability for a street or shade tree
wherever they will grow. The maple is a great favorite, and is a
beautiful tree. It grows rapidly and symmetrically, casts a good shade

in summer, and is pleasant to look upon in winter.   Oaks are admired for their sturdiness, but are slower growing than the maples, and do not form so neat and compact a head, and they do not transplant easily.   The locust presents a rough, crooked trunk and many dead branches, but is valuable for its flowering qualities and quick growth. The linden and tulip tree are of similar character of foliage, are tall and upright, with a bare trunk for some distance and a good, round, close-growing head.   A few evergreens are always desirable about a place to break the monotony of trees of a similar habit of growth and to present something green and snug in winter.   Norway spruces are good while young, but are apt to be ungainly as they grow large. They should be kept well trimmed.   Pines are effective, especially when planted in a group or at some distance from the residence.   It is always desirable to plant a variety of trees, especially upon the home grounds.   A list of some of the best trees is given in succeed-ng chapters.

In laying out a place, trees should be planted in groups, clumps and masses about the borders, with but few single specimens.   A pur-ple beech and other colored or exotic trees look better alone than mixed with other trees.   As a rule, do not mix deciduous and ever-green trees promiscuously in the same clump, but keep them separate. So, also, do not plant oaks and willows together, for they are not only of widely dissimilar habits of growth, but the rapid growing wil-low would soon hide and possibly injure the slower growing oak. Avoid so many trees that the place looks like a forest, but do not plant so few as to give it a barren aspect.   Those of greatly differing characteristics should be somewhat separated.   Planting for color effect in autumn foliage may also be done, and to secure this a care-ful study of the shades of leaf of each variety and species, with the time of their assuming different tints, is necessary.   As a rule, an in-dividual tree takes on the same tint each fall, but this color would probably be made to vary by transplanting the tree to other soil. The autumn color of American foliage is among the brightest in the world, and its effects should be more sought in lawn planting.

CHAPTER II.

FROM NURSERY TO PERMANENT LOCATION.

THE life, shape and usefulness of a tree depend upon the place and method of its early growth, the care it receives in transplanting, the pruning, and the character of the soil in which it grows. Always buy nursery-grown trees, or raise them yourself in nursery rows. This insures tough bark, straight, clean trunk, a tree that is hardy and used to exposure, and above all, one which transplants well. Trees grown in the forest or thicket are long and spindling, unused to the sun and wind, and very difficult to transplant to the open ground. Where pines and some other trees are to be set in large numbers to start a forest, they may be transplanted when very small from a clearing or field where many trees have started up. The important item of first cost is often reduced in this manner. Nearly all trees are better to be transplanted every two or three years in the nursery row before being set in their perminent position. When possible, it is best to set young trees ; they are easier to transplant than large ones, cost less, become established sooner and make a more rapid growth. Young trees, however, are not so easily seen in tall grass, and may often be cut off with the scythe or mowing machine. Therefore, it is necessary to give them good protection until they become large enough to take care of themselves. For this purpose two or three good strong stakes may be driven in about each tree and fastened together with a cord or wire at the top. A box, made of three or four boards nailed at the edges, set up on end around the tree, may also be used, but it has the objection of being a harbor for many kinds of larvæ, and unless fastened to a stake may bend or break the tree or rub the bark. If a rope or wire is to be tied directly around the tree, the bark should be protected by wrapping with cloth or rubber. A piece of old rubber hose to pass the rope through will be found useful. For street purposes, trees cannot be set so young and small as in private grounds or parks. They should be set back from the curbing far enough to be out of reach of horses, and they must also be protected. Wire screens are manufactured for this purpose, and are very useful. They are light, strong, neat, durable, do not harbor insects and will not injure the tree. They can also be taken off or enlarged as the growth of the tree demands. A cheap and durable protector for street trees is shown in the accompanying engraving. It is made of three stakes about which a strand of barb fence wire is wound spirally from bottom to top.

Old trees can be utilized as hitching posts by the use of a three or four foot stick with a snap in one end and which is attached to the tree by a staple. This device is shown in the accompanying engraving (page 8).

Oaks, nut trees, tulip trees and others with a long, straight tap-root must be planted or transplanted when very young. If that is cut off or injured when the tree is several years old, the chances are against the life of the tree. With trees of this class it is often recommended to plant the seeds where the tree is to grow, but young seedlings can be better cared for in a nursery row than when standing alone.

It is sometimes desirable to transplant large trees to new grounds to save years of waiting for young ones to grow. This can be successfully done with proper precautions. First dig the hole where the tree is to be set, and make it large and deep. Then dig a trench around the tree, 10, 15 or 20 feet or more from the trunk, and preserve all roots, bending them in toward the trunk, tying them and covering to prevent drying

Tree Guard.

out. In this way, gather up all that is possible, gradually working nearer the tree and under it until a ball of earth remains six to ten feet in diameter and four to six feet high according to the size of the tree. The tree, roots and ball of earth may then be raised with jack screws and placed on a stone-boat or drag, drawn to its place and let down in a similar manner to which it was raised.   There are vehicles manufactured to raise and carry large trees, and they do it easier and quicker than in the way described.   The hind wheels are high ones, and the axle is raised to carry the roots off the ground.   The tree is bent over, fastened, raised up and balanced, drawn to the desired place, let down and straightened.   After the tree has been let down, lay out the roots in the direction from the trunk in which they were found, and cover them with good, rich earth.   The earth must be

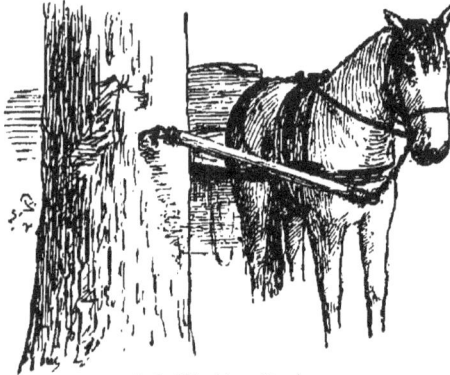

settled down around and under the roots and the ball, and to do this plenty of water and a light tamping must be used.   After the roots are well covered and the ground leveled off around the base, put on a good mulch of hay or straw, and stake and tie the tree so strongly that the wind will not blow it over or out of place.   It is also neces-

Safe Hitching-Device.

sary to prune back the head corresponding to the amount of roots that were removed.   Occasional thorough waterings must be given in dry weather for the first year or two, or until the tree is well estab-lished, when the stakes and stays may likewise be dispensed with.   It is sometimes advisable to dig a trench around the tree a year before it is transplanted in order to develop numerous fibrous roots near the trunk.   Again, the tree may be partly dug and then left until cold weather freezes the ball, when it can be moved easily.   A Norway spruce twenty feet high, transplanted in this way made a good tree, although for several years it apparently did not grow any.

In transplanting a young tree the same general principles are to be observed as with large ones.   Dig the tree so as to save all the roots possible, keep it protected from the sun and wind, and set in large, deep holes, spread out the roots and pack the soil firmly about them, and water if necessary, stake, and mulch as with large trees.   Always set at the same depth that the tree was before digging; if the hole has been dug too deep fill it up to the proper level with good soil. When trees are received from the nursery set them out at once, if the roots are moist and in good condition. If they are dry, either bury them in moist earth for a day or two, or wet them and cover thoroughly with wet hay. With some, such as the Chinese magnolias, it is essential that they be transplanted after growth has begun. If trees are received in the fall, and are not wanted to set until spring, they may be "heeled-in" by covering the roots and part of the trunk with earth. The roots are laid in a furrow made by the plow or spade, and are covered with earth. The tops are laid down so as to nearly recline upon the ground. If the ground is well drained and well prepared, all hardy trees may be set in the fall, but in other cases, set early in the spring. Trees which are set in the fall should usually have a slight mound of earth thrown about them.

In pruning before transplanting, the object should be to reëstab-lish a proper balance between the root and the trunk and branches. Evergreens should have little or no pruning at this time, but deciduous trees will stand a pretty severe cutting, which, however, must be done judiciously. Ordinarily, all side branches may be cut back to one or two buds, and the leader, if slender, should be headed back. Even with trees three or four inches in diameter, like maples, this process is desirable ; the dormant buds push out, and a top is started at any desired hight. With trees of a weeping habit and those grafted high on straight, tall-growing stocks, care must be used not to prune so closely as to injure the graft and start the buds on the original stock. All broken roots should be cut back to fresh wood, and very long, straggling roots may be shortened. Pruning is also necessary with some trees after they are well established, to restrict an unshapely growth in any one direction, to improve the appearance by thinning out branches, or by thickening up, or to train in any given direction or manner. Two rules to be always followed are, (1) never prune without a good reason, and (2) cover all large wounds with paint, wax or some other material. As a general rule, it is well to prune in the spring after growth has commenced, and when far enough along so that the desired effect may be produced. Late winter pruning is safe, however, and labor is usually cheaper at that time. The terminal bud of many trees may be removed, and if done early enough another will form or grow in position to take its place without producing a crooked growth of the branch. Pruning shears or a knife are best to use on young trees, and a sharp saw when large branches are to be removed.

Evergreens may be set during the growing season with good results, providing moist weather follows and great care is exercised to keep the roots moist. Both June and August are favorite months, but June—or when the new growth has just started—is probably the safest month in which to plant in the north. If the growth is heavy and the roots rather small, all the shoots, including the leader, may be headed-in from two to four inches. There is a common notion that cutting off the leader injures an evergreen, but this is a mistake ; a new leader will quickly form. Heading-in the leader tends to make the tree thick and stocky.

Transplanting Large Trees. Ready to Move.

CHAPTER III.

What Trees to Plant.

A LIST of deciduous and evergreen trees is here given for street and shade planting for the northern states. No attempt has been made to describe their botanical characteristics, nor all the varieties and species of a family. A plain statement of the suitableness of each tree for the purpose wanted, with a brief notice of its most prominent features, such as habit of growth, color of foliage or flower, and the soil best adapted to it has been attempted. Many of the newer introductions will undoubtedly prove to be useful for shade and street planting, but in this list only some of the most prominent of those which have been thoroughly tested are inserted.

There is great difference in the quality of shade produced by different trees. Some trees, like the catalpa and basswood, are deep bosomed and dark, and make a dense, cool, damp shade, which is very refreshing at mid-day in summer time. Other trees cast a thin and fleckered shade, which is grateful when one is not tired or overcome by heat. The deeper shades are drowsy ; the lighter ones are playful and restless. These fleckered shadows screen the force of the sun's rays while they do not obscure them, and they seldom afford good retreats. Yet every roadside and pleasure ground should in some measure combine the two. In the cooler days of spring and fall one enjoys the sunshine if it is tempered by thin foliage ; in the summer he hides himself in the dark blankets of maples and basswoods.

Ashes (*Fraxinus*) are not grown to the extent of many other American trees, chiefly because they are usually found in a wet soil, and because the top is thin. They usually grow readily, however, upon upland soils, and are deserving of greater popularity. They do not leaf out until late in the spring. Along the street, ashes usually grow finely and make large handsome trees. They are easy to transplant, and are better adapted for planting along country roads than for city or village streets. They are also attractive and useful about the borders of groves. The white ash is best, with black ash probably second. The European ashes are seldom planted for shade or street purposes.

Bald or deciduous cypress (*Taxodium*), while more particularly a southern tree, will do well in many parts of the north, even as far north as New York city. Being semi-aquatic, it delights in a moist soil, although it often thrives upon dryish uplands. It is not valuable for street planting, but for private grounds or parks it is useful. It is a stately tree of beautiful habit, with small, feathery, light green foliage.

Beeches (*Fagus*) are noted for their rich, glossy foliage and elegant habit, and are among our most magnificent trees. They are very difficult to transplant except when quite small, hence are not used to any great extent for street planting. The American beech (*Fagus ferruginea*) and the European beech (*F. sylvatica*) are large trees, often attaining a hight of 60 to 80 feet. The European beech does not thrive all over the United States as well as the American beech, especially in the interior states. The cut-leaved, purple and weeping beeches are all varieties of the European, and are very ornamental as single specimens on the lawn. Beeches prefer a light, dry soil. Nursery-grown trees should be secured, if possible. For single shade trees in a large lawn or field, the beech is very attractive.

Birches (*Betula*) are not as well adapted for street or shade planting as for ornament. They are too upright and slender in habit of growth, and most varieties have small, fine leaves. *B. Bhojpattra*, (Indian paper birch) is best adapted for shade purposes, having large leaves and being a tree of good size. The native black birch casts a good shade, as do also the larger-growing species, yellow birch and cherry birch. The paper birch (*B. papyrifera*) is valuable as a lawn shade tree in the north. The birches grow best in a light, sandy loam.

Blue gum (*Eucalyptus globulus*) is useful in mild climates for tropical effects in landscape gardening, also for shade. Its wood is valuable for ship building and its leaves for medicinal purposes. It has large leaves six to twelve inches long and of a distinct glaucous hue, unlike any other plant similarly employed. In its native country,

PURPLE-LEAVED BEECH.   *Fagus sylvatica.*

Australia, it often grows to a hight of 300 feet, and although a rapid grower, has hard, durable wood. In California and the southern states it is popular.

Box elder (*Negundo aceroides*, or *Acer Negundo*) is perhaps the most rapid growing native tree which is hardy in the north. The tree is especially valuable in the prairies, where it is planted more largely than any other tree. It is extremely hardy. The leaves resemble those of the ash, and it is often called ash-leaved maple. It is adapted to streets, and is sometimes used in lawns, but it is a cheap-looking tree, and there is danger of over planting it.

Catalpas are rapid-growing, tropical looking trees, that are often grown for their dense shade and their flowers, which are borne in profusion in July, when other trees are not in bloom. One of the best is *C. bignonioides* (syn. *C. syringæfolia*), a native of the south and a rapid-growing, showy tree. *C. speciosa* is a finer, hardier tree and better adapted for extensive planting in the north. Catalpas thrive in almost any rich, porous soil.

Cherries, especially the native black or rum cherry (*Prunus serotina*) are useful for street and shade trees. The wild black cherry is one of the most attractive native trees, attaining a large size and growing very rapidly. The sweet or mazzard cherry (*Prunus Avium*) has run wild in many parts of the east. These mazzards often make trees two feet and more in diameter, and are hardy and valuable. The sour or morello cherries are too low for street trees.

Elm (*Ulmus*) ranks with the oak and maple in stateliness of growth. As the oak is called the king of the forest, the elm might be styled the queen of the field. The tall, broadly waving branches cover a large area, and trees 100 feet high are common. The American or weeping elm (*Ulmus Americana*) is probably the handsomest of our American trees. The majority of ornamental kinds are varieties of the English elm (*U. campestris*). The elm does well in a variety of soils, but a moist, deep rich one is best suited to its rapid growth and perfect development. There are three species native to the eastern states, of which the best is the common American elm mentioned above. This runs into a great variety of forms, which, however, do not always indicate their characteristic features when young. The so-called water elm, swamp elm, rock elm, white elm, etc., are all mere forms of this one species. The cork elm (*U. racemosa*) is a curious slow-growing tree with stiff corky, winged branches, rarely planted. The slippery elm (*U. fulva*) is seldom planted because of its stiff open habit, slow growth and dull foliage.

Fruit trees are less desirable for street or road planting than nut trees (which see). Plum and peach trees are too small for shade purposes along a road or street, and also too easily broken, but apple, pear and cherry trees attain good size. The apple is inclined to grow low and spreading, and the branches will be in the way as the tree grows old. Pears are slow-growing trees, but sometimes many reach very large size. Many fruit trees, and especially apples, have too rough leaves for road-side planting. Such leaves hold the dust. Fruit trees rarely bear well upon the street, because of neglect.

Ginkgo, or maiden-hair tree (*Salisburia*, or *Ginkgo adiantifolia*) is a native of Japan and notable as an oddity. It combines the characteristics of the conifer and the deciduous tree in its fruit and habit. The tree is of medium size, rapid growth, with beautiful foliage of the maiden-hair fern type. It thrives well in any good soil, and is adapted for ornamental purposes rather than for extensive planting. As it grows tall and straight, it can be recommended for narrow streets.

Horse chestnuts and buckeyes (*Æsculus*) are trees of magnificent foliage, large growth and fine habit. They are rapid growers, and in the spring are beautiful with the abundance of large panicles of white or yellow or reddish flowers. The heads are very round and compact, and often form such a dense shade that grass will not grow beneath. They are a heavy tree on the lawn, and their fruit, which ripens in the fall, is so much in demand by small boys for missiles that the branches in consequence are often broken and injured. The Ohio buckeye (*Æsculus glabra*) is a well known tree of the western states. It has yellow flowers, and blooms before the others. The double white-flowered horse chestnut is a superb variety, with double flowers and little or no fruit. Horse chestnuts do well in most soils, but the more loamy it is the better. The common or Old World horse chestnut is most too stiff and formal for extensive planting.

Kentucky coffee tree (*Gymnocladus Canadensis*) is a fine native tree of medium size, resembling a locust or acacia in foliage. It is of rapid upright growth, has very rough bark, stiff blunt shoots and feathery foliage of a bluish green color. It is very hardy and thrives best in a deep, rich soil. It is easy to transplant if nursery-grown stock is obtained, and should be planted more extensively. Its shade is thin but grateful. It bears very large bean-like pods. Some trees are entirely staminate-flowered and produce no seeds.

Larch is a name applied to the various species of the genus *Larix*, which is one of the pine or cone-bearing family, although the larches, like the bald cypress, are deciduous trees. The native larch or tamarack (*Larix Americana*) inhabits swamps, but in cultivation it thrives in most any moist soil. The European larch (*Larix Europæa*) is a broader and shorter tree, which thrives upon dry soils. Both species are occasionally seen on road-sides, where they make good effects. A clump of the tamarack in low places on the road preserves the natural look of the country and makes a pleasing contrast with the commoner road trees.

Linden, basswood or lime tree (*Tilia*) is a beautiful genus, and deserves more attention than it receives. The trees are of rapid growth, beautiful habit and useful for soft timber as well as for ornament. The flowers, which are borne in profusion in late spring or early summer, are very sweet and handsome and attractive to bees. The ornamental-leaved and ornamental-barked varieties are all forms of *Tilia Europæa*, the European linden. These named varieties are very useful for home grounds, but for street trees the common wild basswood (*T. Americana*) is probably the best, although the European linden is very generally planted in the east. There is little difference in foliage between the two. All the lindens or basswoods cast deep and quiet shade, and are valuable for this reason. They thrive upon most soils, although the moister ones are better.

Locust or acacia (*Robinia*) makes a good shade tree, and is very handsome when in bloom. The common or flowering locust (*R. Pseudacacia*) is valuable for its timber, and is a rapid-growing tree of good form. It has large panicles of white or yellowish blossoms, which are very sweet scented. The rose or moss locust (*R. hispida*) is a shrub of spreading, irregular growth, and bears elegant clusters of rose-colored flowers during June and at intervals throughout the season. The honey locust, or three-thorned acacia (*Gleditschia triacanthos*), belongs, like the last, to the pea family, but the flowers are comparatively inconspicuous. The tree is a very open, straggling and picturesque grower, and the long curled or crooked pods are curious. The tree is densely armed with spines or thorns when it attains some size. The honey locust is valuable for roadside planting. It is always attractive. There is a thornless variety of the honey locust. The shade of both the locusts is rather sparse. Any soil not too moist will do for locusts. They are transplanted readily and grow with great vigor, making valuable posts in a short time.

HONEY OR THORN LOCUST.

Magnolias for lawn effects, especially in groups, are not excelled. They are handsome, tall trees of stately form and splendor of growth. They have beautiful flowers that are borne in abundance in early spring and large, rich, glossy foliage. They are better adapted for the lawn than for the street. They should always be transplanted in the spring, and the Chinese species and varieties at the period when they are coming into growth. Use great care in their removal, pre-serve all the fibrous roots possible and protect them from the wind and sun. There are many kinds of magnolias in cultivation, few of which are adapted for shade or street purposes in the north. The hardiest one is the cucumber tree. The swamp laurel or sweet bay (*Magnolia glauca*) is also hardy as far north as New England, but it rarely attains sufficient size in this latitude to make it valuable for shade. The umbrella tree (*M. tripetala*) is useful for a lawn shade tree in the north. *M. speciosa, M. Soulangeana* and *M. Lennei* are among the best and hardiest foreign kinds. The cucumber tree (*M. acuminata*) is a tall-growing, beautiful pyramidal tree. The leaves are long and bluish-green, the flowers yellow, tinged with bluish purple and the fruit when green resembles a cucumber. This grows wild from New York southward. The great-leaved magnolia (*M. macrophylla*) is a medium sized tree with leaves nearly two feet long and flowers of immense size; not hardy north. Thurber's Japan magnolia (*M. Kobus*) is one of the tallest of the Chinese magnolias, and has fragrant blush-white flowers. Magnolias will grow in any good soil, but one that is rich, warm and dry is best.

Maples (*Acer*) are among our most valuable and ornamental trees. They are of large growth, broad rather than high, but often attaining a hight of seventy-five feet or more. They are easily transplanted, and do well in a variety of soils. The head is symmetrical and spread-ing, casting a good shade, but not too dense. In rows along streets they are especially fine, and being vigorous growers and free from disease, they are deservedly popular for this purpose. The white, silver-leaved or soft maple (*A. dasycarpum*) is a rapid growing tree, and has foliage bright green above and silvery beneath. The sugar, hard or rock maple (*A. saccharinum*) is of slower growth than the former, but more symmetrical and desirable. The red or swamp maple (*A. rubrum*) is a small tree, with small, rather light foliage, which is brilliant in autumn. The red flowers in very early spring are very attractive. It deserves more attention as a road tree, especially in the lower places. *Acer campestre*, the European maple, is a desir-able small tree for the middle latitudes. The Norway maple (*A. platanoides*) is a low-headed very dense tree, which casts the deepest shade of all the maples. As a lawn shade tree it is unexcelled, but the head is too low for street planting. The sycamore maple (*A. Pseudo-platanus*) is a rather small tree, with dense, dark foliage, and is useful for both lawns and streets. In the northern states it is often injured by the winters. There are a number of species and varieties with ornamental foliage, which makes them very useful as single spec-imens. Most of these, however, are large shrubs or small trees. Wier's cut-leaved silver maple is a form of *Acer dasycarpum.* The Japan maples are large, handsome shrubs. Maples like a deep, loamy, well-drained soil.

Mountain ash is a pleasant tree of rapid growth, symmetrical form and long, slender branches. The fruit is bright red, borne in bunches and often hangs on until winter, giving the tree a very distinct and handsome appearance. Borers are the great enemy and the tree is not planted largely on this account. They can be dug out or killed by thrusting a sharp wire into their burrows. The common mountain ash is the European rowan (*Pyrus Aucuparia*), although it is often sold for the American species. The American mountain ash, while an excellent tree, is not so much sold by nurserymen as the other. These trees are not ashes, but are closely related to apples and pears. They are useful for lawn shade trees and bowers. The mountain ash grows well in any good soil.

Mulberries of various kinds are useful for shade, especially the com-mon large half-wild forms of the white mulberry (*Morus alba*). This species attains to a large size, often nearly or quite two feet in dia-meter and resembles a much exaggerated high old apple tree. The leaves are glossy and attractive. The fruit varies from almost white to violet and purple-black, and is much sought by poultry and birds. The Russian mulberry is a form of this species. It is usually a low tree or large shrub, very hardy, but valueless for fruit. The true black mulberry (*M. nigra*) is very little known in America, and is not hardy in the northernmost states. The native or red mulberry

(*M. rubra*) is a hardy and most robust tree of medium size in the north, but reaching a hight of seventy feet in the south. It should be better known as a shade and street tree.

Nettle tree (*Celtis occidentalis*) is a native tree of moderate size and fine, spreading habit, with the look of an elm. It has numerous slender branches and thick, rough bark. The leaves are about the same size as those of the apple, but more pointed, and of a bright shiny green. It does well in any good soil, and should be better known, both on the street and lawn.

Nut trees are desirable for street planting, but when the nuts begin to ripen boys are continually climbing the trees or throwing sticks and stones to knock them off, and the trees are soon injured, broken and otherwise disfigured. On the lawn they are usually dirty trees, but they give good shade and are handsome specimens. The hickories (*Carya*) are slow-growing trees of handsome, upright habit. The common white or shag-bark hickory is perhaps the best for shade, although the bitter-nut is a unique, feathery tree, and worthy of extensive planting. Hickories usually transplant with difficulty, unless they are nursery-grown or have had the tap-root cut a year or two before removal. The pecan (*C. olivæformis*) is not hardy in the north. The walnut is the most valuable of all our native trees for timber. The black walnut (*Juglans nigra*) is easily grown, and if carefully handled can be transplanted with safety. It is a noble roadside tree. The butternut (*J. cinerea*) resembles the black walnut in growth, but has smooth and less rugged branches and usually a thicker top. It is also an ideal street tree. The English walnut, (*J. regia*) is not hardy in the north. The chestnut (*Castanea Americana*) would be a valuable tree could the boys be kept from injuring it. It grows rapidly and makes a handsome top, which is broad and dense. The native chestnut is the best tree for street and shade planting, and the nuts, while small, are better in quality than those of the European species. Nut trees grow well in any fertile soil. Chestnuts do better in a lighter soil than most of the others.

Oaks (*Quercus*) are the grandest and sturdiest of all our trees. They are branching and spreading in their habit of growth, and old trees are very picturesque and bold. The leaves of many species cling to the branches with great tenacity, and many remain on all win-

ter and even after the young growth has started in the spring. They
turn to red, brown, scarlet or purple in the fall, and sometimes remain colored until they fall. Being of slow growth and hard to transplant after they have attained any size, oaks are not adapted for
street planting unless nursery grown. The American white oak
(*Quercus alba*) is one of the finest of our native trees in habit, although
other oaks excel it in foliage. The scarlet oak (*Q. coccinea*) and the

A LIVE OAK COVERED WITH SPANISH MOSS.

black oak (*Q. coccinea* var. *tinctoria*) are of more rapid growth than
the white oak. These are especially desirable for their bright autumn
foliage. The leaves of these species are very deeply cut, and the
acorns are small, with a deep cup. The red oak (*Q. rubra*) is of
large size and rapid growth, with handsome foliage. The leaves are
less deeply cut than in the last two, and the acorn is large, with a very
shallow cup. The tree is a very open grower. The live oak (*Q.*

virens) is a conspicuous and useful evergreen tree in the southern United States. It is shown in the cut. Other northern oaks of great merit are the burr oak (*Q. macrocarpa*) and the swamp white oak (*Q. bicolor*). Two or three chestnut-like oaks, of which the chief is *Quercus Prinus*, are also desirable trees for shade. The willow oak (*Q. Phellos*) is prized in the middle states. The English oak (*Q. Robur* or *Q. pedunculata*) is a very slow-growing tree in America, and is rarely planted for shade purposes.

Pepperidge or gum tree (*Nyssa multiflora*) is useful as an oddity. It reaches 30 to 50 feet in hight and is chiefly grown because of its foliage, which turns to a bright crimson in autumn, and its picturesque habit. The trees present a most curious, deflected habit of growth, which makes them exceedingly striking in the landscape. Every lawn should have a specimen. The pepperidge is used to best advantage, however, when planted in small groups. A low, damp, moist situation, as a swamp, is best suited to it, although it often thrives upon uplands.

Poplars (*Populus*) are very useful as ornamental, quick-growing trees. They are of upright growth, and thrive in a variety of soils, but do best in a moist one. The white or silver poplar (*P. alba*) is a tree of wonderfully rapid growth and wide spreading habit, but it sprouts so badly that it is in great disfavor. The thick leaves are very white beneath and maple-like in shape, and for this reason the tree is sometimes called silver maple. The Lombardy poplar (*P. fastigiata* or *dilatata*) is useful in landscape gardening to break the somber effect of most other trees. It is a tall-growing species, often growing as high as 150 feet. It is a tree that is commonly over-planted, however. A row of Lombardies along a street is exceedingly formal and the shade is not great. The tree sprouts badly along highways, and is apt to be a nuisance. In the northern states the old trees are often injured by the winters. A clump of Lombardies here and there upon some conspicuous site makes a very bold and picturesque object. The common cottonwood or Canadian poplar (*P. monilifera*) of our central and western states is a tall-growing tree, and delights in moist soil, especially along streams. It is one of the most useful of all the poplars. The common aspen (*P. tremuloides*) is a neglected tree. The bright, dangling leaves make it a gay tree. It does not attain to a large size and is therefore not adapted to roadside use. The large-toothed poplar (*P. grandidentata*) is a larger growing tree of a heavier cast. The balm of Gilead (*P. candicans*) is one of the freest-growing and most useful members of the genus. The poplars should never be planted heavily, but when used with discretion they are among the most useful trees.

Red bud or Judas tree (*Cercis Canadensis*) is a medium sized, ornamental tree, that has reddish-purple flowers, which cover the branches before the leaves appear. The head is of irregular, roundish shape, and the round leaves glossy green above and grayish green beneath. The tree does best in a deep, friable, sandy loam. It is not large enough for street planting, but can be used for shade in small lawns.

Sweet gum or bilsted (*Liquidambar Styraciflua*) makes a magnificent tree in the south and middle states, often attaining a hight of 100 feet and four or five feet in diameter. North of Pennsylvania it is not usually hardy without protection. The head is roundish or tapering, and the tree is beautiful at all stages of its growth, but especially in the autumn when the leaves, which have been a beautiful glossy green all summer, turn to a purplish crimson. A moist, loamy soil is best suited to it. It has the habit of a cottonwood with the leaves of a maple.

Sycamore, buttonwood or American plane tree (*Platanus occidentalis*) is a tree of large growth and beautiful form, well adapted to plant along country roads and in lawns or parks. It requires a rich, deep, moist soil, and does best when its roots can reach the water. The oriental or common plane (*P. orientalis*) is a fine species, growing 60 to 80 feet high, but is not hardy in the north. The plane tree is conspicuous for its long bare white arms and its curious leaves and balls of fruit. It is easily transplanted, and grows rapidly. It is not properly appreciated.

Tulip tree or whitewood (*Liriodendron Tulipifera*) is a magnificent native tree of tall, pyramidal habit. The flowers, which are borne in abundance as the tree attains age, resemble tulips, and the leaves are broad, glossy and of a bright-green color. The trees are difficult to transplant except when very small, unless they are nursery-grown. For the lawn it is one of the best of the trees of this habit. A deep, loamy soil is best suited to its growth.

Willows (*Salix*) are very rapid-growing trees, hardy and adapted to a great variety of soils and purposes. The white willow, which came from Europe, is popular where large trees are wanted in a short time. It is transplanted very readily, and even willow posts, if fresh, driven in the ground will root and grow. The willow prefers a moist soil. This and similar willows are usually not desirable for lawn trees, because they drop their leaves and twigs so easily. Along water courses and in low places and near springs they are often very desirable. There are many species and varieties of both upright and drooping growth. A single specimen of *Salix Babylonica*, the common weeping willow, makes a very handsome tree. The native peach willow (*S. amygdaloides*) is one of the neatest of the native small trees, and should be better known. The black willow (*S. nigra*), also a native, is a useful tree in low places. The willows are not adapted for city streets, as the soil is usually too dry.

Yellow wood (*Cladrastis tinctoria* syn. *Virgilia lutea*) is one of the finest of our American trees. It is a moderate grower, with a broadly rounded head, leaves of a light-green color, which turn yellow in autumn. The flowers are white and appear in June in great profusion in long, drooping recemes. It is one of our finest hardy flowering trees, and does well in any soil or situation.

Conifer.ɛ or Evergreens.

The arbor-vitæ (*Thuya*) is more used as a hedge plant than for a shade tree because of its small size and close, upright growth. The white cedar (*T. gigantea*) a native of the northwest, often reaches a hight of 150 feet, and is a graceful, stately tree, suitable for a position on any large lawn. The American or western arbor-vitæ (*T. occiden-talis*) and its varieties, are more hardy than its eastern or oriental neighbor (*T. orientalis*) which, although a handsome, and more ornamental tree, is not hardy enough for this climate. The leaves of the arbor-vitæ are small and scale-like and the habit of growth close, compact and upright. It is useful for single specimens and windbreaks.

Cypresses (*Cupressus*) make elegant, large trees where they are hardy, but they will not stand our severe winters. Lawson's cypress (*C. Lawsoniana*) forms a very large, handsome tree in California, but is not more than half hardy in New York. It has elegant drooping branches and dark, glossy green leaves, tinged with a glaucous hue. The cypress will grow in any good garden soil, but one that is deep, moist and rich suits it best.

Pines (*Pinus*) are of the greatest value in landscape and ornamental gardening, as well as for practical utility for shade trees. They are rapid growers, and nearly all kinds of soils and latitudes are suited to their growth. The white pine (*P. Strobus*), perhaps the best of our native pines, has light, delicate and silvery green foliage, and flourishes in the poorest, light, sandy soil. The red or Norway pine (*P. resinosa*) rivals the white pine for ornamental planting. It is of rapid growth, good form, long life, and perfectly hardy when once established. The Soctch pine (*P. sylvestris*) is a quick-growing tree and especially useful for hedges. Austrian pine (*P. Austriaca*) is one of the most rapid-growing of all the species, and is much planted in this country for shade and shelter. It is a coarse species, and should not be planted near the residence. The heavy wooded or bull pine (*P. ponderosa*), a Rocky mountain species, grows well in very light, dry soil, and reaches great size, but is not hardy in the northernmost states. It does not do so well in a moist atmosphere. The Monterey pine (*P. insignis*) is one of the handsomest of all. It is, however, only hardy enough to bear mild winters.

Spruces, firs and hemlocks belong to the cone family. The leaves of the spruces and hemlocks are needle shaped, and grow all about the shoots ; those of the firs are flat and two ranked. The limbs are slender, grow nearly horizontal from the trunk, and are inclined somewhat upwards or downwards in the different species. Trees of this genus are of great value for hedges, windbreaks and as ornamental specimens, but are not useful as shade trees until they reach considerable size. The Norway spruce (*Picea excelsa*) is very hardy, easily transplanted, but a coarse, unsightly tree when it reaches age, unless the tips of the shoots have been cut in from year to year. Our native hemlock (*Tsuga Canadensis*) is a very beautiful and graceful tree, and handsome on the lawn. It has delicate dark foliage somewhat like the yew, and is distinct from all other trees. It should be planted more extensively upon private grounds. The white spruce (*Picea alba*) is a native tree of medium size, very hardy and valuable. The

Colorado blue spruce (*Picea pungens*) is one of the most beautiful and valuable of all the spruces. The foliage is of a rich blue or sage color and the tree is very hardy, and similar in habit and growth to the white spruce. The firs are not as popular as the spruces or hemlocks, but are very handsome trees. Nordmand's silver fir (*Abies Nordmanniana*) is one of the most beautiful species, of symmetrical form, vigorous and quite hardy. A moist, rich soil is best suited to the growth of hemlocks, firs and spruces. They are not well adapted for shade purposes alone, but as single specimens and for groups and shelter belts they are indispensable.

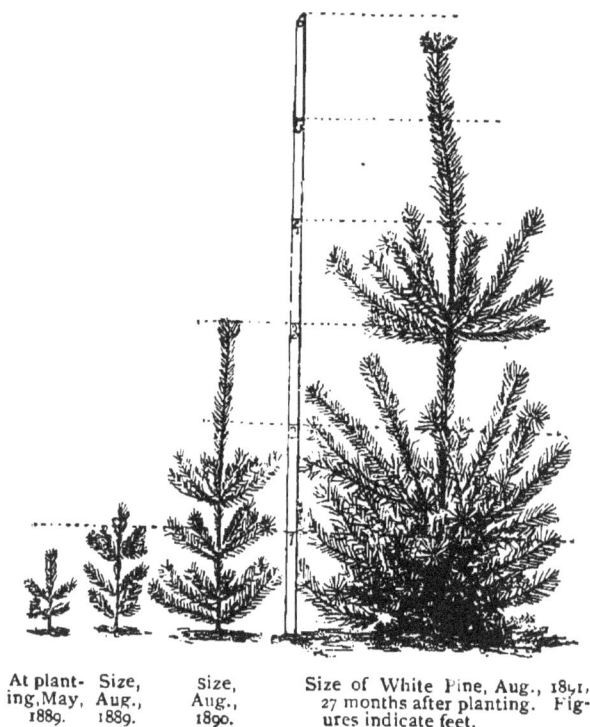

At plant- Size,   Size,   Size of White Pine, Aug., 1891,
ing,May, Aug.,   Aug.,   27 months after planting. Fig-
1889.   1889.   1890.   ures indicate feet.

CHAPTER IV.

HARDY TREES FOR PARTICULAR PURPOSES.

The following lists of trees, largely native species, which are adapted to particular uses in the north, is part of a catalogue prepared in 1887 by L. H. Bailey. The trees are arranged somewhat in the order of their merit. The species of each genus are strictly so arranged so far as possible.

I. Trees for Shelter Belts.

White pine, *Pinus Strobus.*

Austrian pine, *P. Austriaca.*

Scotch pine, *P. sylvestris.*

Red pine, *P. resinosa.*

Norway spruce, *Picea excelsa.*

Any of the rapidly growing, native forest trees, especially :

American elm, *Ulmus Americana.*

Sugar maple, *Acer saccharinum* and var. *nigrum.*

Basswood, *Tilia Americana.*

Cottonwood, *Populus monilifera.*

Balsam poplar, *P. balsamifera.*

Wild black cherry, *Prunus serotina.*

II. Trees for Groups or Single Specimens.

A. *Deciduous trees.*

Norway maple, *Acer platanoides.*

One of the finest trees for single lawn specimens, especially in tranquil scenes. It droops too much for roadside planting.

Black sugar maple, *A. saccharinum* var. *nigrum.*

Darker and softer in aspect than the ordinary sugar maple.

Sugar maple, *A. saccharinum.*

This and the last are the best roadside trees.

Wier's cut-leaved silver maple, *A. dasycarpum,* hort. var.

Light and graceful. Especially desirable for pleasure grounds.

Silver maple, *A. dasycarpum.*

> Desirable for water-courses and for grouping. Succeeds on both wet and dry lands.

Red, soft, or swamp maple, *A. rubrum.*

> Valuable for its spring and autumn colors, and for variety in grouping.

Sycamore maple, *A. Pseudo-platanus.*

> A slow grower, to be used mostly as single specimens.

American elm, *Ulmus Americana.*

> One of the most graceful and variable of trees ; useful for many purposes.

Cork elm, *U. racemosa.*

> Softer in aspect than the last, and more picturesque in winter. Slow grower.

Red, or slippery elm, *U. fulva.*

> Occasionally useful in a group or shelter belt. A stiff and straggling grower.

European silver basswood, *Tilia argentea* and varieties. (*T. alba.*)

> Very handsome. Leaves silvery white beneath. Among others, is a weeping variety.

American basswood, *Tilia Americana.*

> Very valuable for single trees on large lawns, or for roadsides.

European basswood, *T. Europæa* and varieties.

Tulip tree or whitewood, *Liriodendron tulipifera.*

> Unique in foliage and flower. Should be in every collection.

Cucumber tree, *Magnolia acuminata.*

> Not reliable north of Buffalo and Detroit. Handsome.

Yellow wood, or Virgilia, *Cladrastis tinctoria.*

> The finest hardy flowering tree.

Swamp white oak, *Quercus bicolor.*

> A very desirable tree, usually neglected. Very picturesque in winter. The oaks are slow growers and transplant with difficulty. Natural specimens are most valuable A large oak, well grown, is one of the grandest of trees.

Burr oak, *Q. macrocarpa.*

Chestnut oak, *Q. Prinus,* and especially the common var. *acuminata,* or *Q. Muhlenbergii.*

White oak, *Q. alba.*

Shingle oak, *Q. imbricaria.*

Scarlet oak, *Q. coccinea.*

> This and the next two are glossy-leaved, and are desirable for gay scenes.

Black oak, *Q. coccinea* var. *tinctoria.*

Red oak, *Q. rubra.*

Pepperidge or gum-tree, *Nyssa multiflora.*

One of the oddest and most picturesque of our native trees.   Especially attractive in winter.   Most suitable for low lands.

Horse chestnut, *Æsculus Hippocastanum.*

Useful for single specimens.

Showy catalpa, *Catalpa speciosa.*

Very dark, soft foliaged tree of small to medium size.   Showy in flower.   To be used as single specimens.

Smaller catalpa, *C. bignonioides.*

Less showy than the last, blooming a week or two later.   Less hardy.

Black ash, *Fraxinus sambucifolia.*

One of the best of the light leaved trees.   Does well on dry soils.   Not appreciated.

White ash, *F. Americana.*

Kentucky coffee-tree, *Gymnocladus Canadensis.*

Light and graceful.   Unique in winter.

Bitter-nut, *Carya amara.*

Much like black ash in aspect.   Not appreciated.

Hickory, *C. alba.*

Useful in remote groups or belts.

Cut-leaved weeping birch, *Betula alba,* hort. var.

The finest of gay trees.

Cut-leaved birch, *B. alba,* hort. var.

European birch, *B. alba.*

American white birch, *B. alba,* var. *populifolia.*

Paper or Canoe birch, *B. papyrifera.*

Purple birch, *B. alba,* hort. var.

Cherry birch, *B. lenta.*

Well grown specimens resemble the sweet cherry.   Both these and the next make attractive light leaved trees.   They are not appreciated.

Yellow birch, *B. lutea.*

Aspen, *Populus tremuloides.*

Very valuable when nicely grown.   Too much neglected.   Most of the poplars are suitable for pleasure grounds

Larged-toothed aspen, *P. grandidentata.*

Unique in summer color.   Heavier in aspect than the last.   Old trees become ragged.

Weeping poplar, *P. grandidentata,* hort. *pendula.*

An odd, small tree, suitable for small places.

Cottonwood, *P. monilifera.*

Desirable in a group or near water.   The staminate specimens, only, should he planted, if possible.

Balm of Gilead, *P. balsamifera,* var. *candicans.*

Desirable in remote groups or belts.   Foliage not pleasant in color.

Lombardy poplar, *P. dilatata.*

Desirable for certain purposes, but too much in use.   It is apt to be short lived in the north.

Bolle's poplar, *P. Bolleana.*

Habit much like the Lombardy.   Leaves curiously lobed, very white beneath, making a pleasant contrast.

Locust, *Robinia Pseudacacia.*

Should be planted at some distance from the dwelling.   Useful in grouping. Attractive in flower.   Handsome as single specimens when young.

Honey locust, *Gleditschia triacanthos.*

Like the last, this should be planted rather remote from the residence, or near the borders.   The foliage of both is light.

Beech, *Fagus ferruginea.*

Specimens which are symmetrically developed are among our best lawn trees. Picturesque in winter.

Chestnut, *Castanea vesca* and var. *Americana.*

Plane or Buttonwood, *Platanus occidentalis.*

Young or middle-aged trees are soft and pleasant in aspect, but they soon become thin and ragged below.   Most desirable in belts.   Unique in winter.

Sassafras, *Sassafras officinale.*

Suitable in the borders of groups or for single specimens.   Peculiar in winter. Too much neglected

Maiden-hair tree, *Solisburia adiantifolia.*

Very odd and striking.   To be used for single specimens.

Rowan or European mountain ash, *Pyrus Aucuparia.*

Weeping willow, *Salix Babylonica.*

To be planted sparingly, preferably near water.   The sort known as the Wisconsin weeping willow appears to be much hardier than this type.

White willow, *S. alba,* and various varieties, one of which is the Golden willow.

May be used sparingly.

Wild black cherry, *Prunus serotina.*

Nettle-tree, *Celtis occidentalis.*

Box elder, *Negundo aceroides.*

European larch, *Larix Europæa.*

American larch or tamarack, *L. Americana.*

Bald cypress, *Taxodium distichum.*

Not entirely hardy in New York. Becomes scraggly after fifteen or twenty years.

Butternut, *Juglans cinerea.*

Walnut, *J. nigra.*

Ailantus, *Ailantus glandulosus.*

A rapid grower, with large pinnate leaves. The staminate plant possesses a disagreeable oder when it flowers. Suckers badly.

B. Coniferous Evergreens.

Norway spruce, *Picea excelsa.*

Loses much of its peculiar beauty when thirty or fifty years of age.

White spruce, *P. alba.*

One of the finest of the spruces. A more compact grower than the last, and not so coarse. Grows slowly.

Oriental spruce, *P. orientalis.*

Especially valuable from its habit of holding its lowest limbs. Grows slowly.

Blue fir, *P. pungens.*

In color probably the finest of the conifers. Grows slowly. Varies in blueness.

Nordmann's fir. *Abies Nordmanniana.*

Balsam fir, *A. balsamea.*

Loses its beauty in fifteen or twenty years.

Hemlock spruce, *Tsuga Canadensis.*

Young and well grown specimens are the most graceful of our evergreens. If given some protection from the sun it does better. Should theretore be planted near large trees.

Arbor-vitæ, *Thuja occidentalis.*

Becomes unattractive after ten or fifteen years, especially on poor soils.

Cembran pine, *Pinus Cembra.*

A very fine slow growing tree. The only pine suitable for small places.

White pine, *P. Strobus.*

Scotch pine, *P. syvestris.*

Red pine, *P. resinosa.*

Valuable in groups and belts. Not sufficiently known. Usually called "Norway pine."

Scrub pine, *P. Banksiana.*

A small tree. Picturesque.

Red cedar, *Juniperus Virginiana.*

A. Deciduous Trees.

CHAPTER V.

SHADE TREES IN CITY STREETS.

By Wm. McMillan. *

S HADE trees along the borders of the streets were at one time
a distinguishing feature of American cities. But this distinc-
tion is not now as marked as it formerly was. Our example
has been largely copied in the suburban districts of nearly all
the rapidly growing cities of Europe. In many of them the street
planting has been more thorough, systematic and successful, than in
any American city, Washington excepted. In the chief business sections of every large city the conditions are fatal to street trees.
Where the population is dense, the traffic large, the atmosphere
smoky and dusty, and the sidewalk borders almost impervious to
water, trees cannot thrive or long endure. Even those which may
survive for a time, as if defying the common fate, are cut down to
give more room for the daily business of the streets. Thus even in
the most fashionable residence sections of every large city fine avenues of trees are now rarely seen, except in the newer outskirts and
adjacent suburbs. But away from the larger marts of trade our early
pre-eminence in street planting is probably still maintained. Wherever the conditions are favorable our streets are embowered in foliage,
and on every new street that is opened the planting usually keeps
pace with the other improvements. We have many fine examples
of this in all sections of the country, but we need not look far away
for a model. This city of Rochester is as good an example as any.
Perhaps no city in the country has had equal advantages in soil, subsoil, natural drainage, tree supply, good example and public spirit.
If in summer we take a bird's-eye-view of the town from the outlook
pavilion in Highland Park, the houses seem to be nearly hidden by

* An address given before the annual meeting of the Western New York Horticultural
Society, at Rochester, January, 1893.

the trees. The section where trees have given way to business is probably smaller than in any other city of its size.

In nearly all our cities, street trees are set out and cared for solely by the owners or occupiers of the abutting property. Each man plants, or not, according to his own taste or interest in the matter. This involves much diversity, inequality and incongruity in the selection of species, in age and size, in the distance from the curb line and from each other. Uniformity in these respects can only be obtained for any given stretch of street where the work is done by municipal authority. This method has been eminently successful in the leading cities of Europe, and in Washington, where nearly every street that has been opened and graded has been systematically planted under the central authority of a special commission. But the general custom in this country indicates that the American temper resents official control of such matters. It is probable that the Washington experiment will long remain the solitary exception on a large scale. In any other city it would be difficult to organize a planting commision with the same practical knowledge and singleness of purpose among its members, the same prestige of authority behind it, and the same respect for the law in the community. We have several examples on a small scale in cities where park commissioners have had exclusive control of the planting on certain avenues used as park-ways. Where these have been of extra width, offering favorable conditions of soil and situation, the planting as a rule has been fairly successful, but where the ordinary conditions of street usage and sidewalk space have been encountered failure is common. This is due not only to the difficulty of securing favorable conditions of soil, moisture, and space, but also to the lack of active co-operation, and constant vigilance in protecting the young trees from damage, which can only be secured where each abutting property owner not only feels a personal interest in his own street frontage, but entire responsibility for it. You may think such aid would be always available, but really effective assistance is rare. Many openly condemn the scheme of the planting, and ignore the benefits they receive by it, apparently just because it cost them nothing. The rule that requires every man to sweep his own sidewalk is equally applicable to the care of the trees.

Few realize the constant liability to damage and destruction to which youug street trees are exposed. Some idea of it may be gained by inspecting the trees on any given street, and noting how few show no sign of stunted growth, scarred trunk, mutilated top, or blemish of some kind. The most common damage is the gnawing of the bark by horses, or of the branches if within reach, but up to a certain age mischievous boys are far more destructive. If the sapling gets safely out of its swaddling clothes it is next attacked at the roots by trenches for sewers, gas pipes, water pipes, and electric cables, or by changes of line or grade in laying curbstones or flagging. In later years the largest limbs will be cut off to open a view, or the top mutilated by telegraph line-men and their wires. Again, under ordinary conditions the trees suffer constantly from lack of moisture, because the pavement or the beaten ground sheds most of the rainfall; from lack of food, because the roots cannot penetrate the hardened subsoil ; from poison by gas, because the small service pipes soon become rusted through ; and from want of air, because the soot and dust of the city stops up the pores of the leaves. The unhealthy condition resulting from these and other causes invites grubs and borers, bugs and caterpillars, scale, spider and fungus, all in great profusion. In the streets these insect pests are safe from their natural enemies—the birds—and from the poisonous spray of the gardener's syringe.

This brings up the important questions : What kinds of trees are best suited to withstand these untoward conditions ? What methods of treatment are most likely to protect them from injury and disease, and secure healthy growth and long life ? Of course no list can be equally suitable to all localities, and no rules equally applicable every-where, or even generally acceptable among experts. But the good and the bad points of certain standard kinds may be broadly noted, and good average conditions of cultivation indicated. The trees most commonly used are probably the best under average conditions. Their presence in every town and their general appearance indicate special adaptability. "Nothing succeeds like success, " and the points contributing most to this success are, ease of propagation, cheapness of nursery culture, quickness of early growth, endurance under careless transplanting, prominence of good looks and absence of bad habits, the ability to pick up a living on a scanty diet, and patience under abuse of every sort.

The silver maple has apparently fulfilled these requirements, all in all, more generally than any other tree, for it is the one most generally planted. If in rich soil with sufficient moisture, its growth is too rank for strength ; it does not hold its head erect, and its long, slender branches do not bear well the strain of high winds.

The white elm is next in popular favor. In many of its traits it is far superior to the white maple. Its tall trunk, lofty head, wide sweep of the branches and pendent spray make it an ideal street tree —beautiful to the eye, and giving a generous proportion of light in its shadow. Wherever the subsoil is loamy or porous its fibrous roots extend deeply as well as widely, insuring a thrifty growth. It grows well in stiff clay, if moist enough, but under a sidewalk that is rarely the case. Dry clay is as unfavorable to any root growth as the most sterile hardpan.

The European elms also do well as street trees. The broader leaved species, known as the Scotch elm, has usually a higher and broader head than the English elm, but neither has the graceful form nor the open shade of the American tree.

The sugar maple is largely planted in the cities of the north-eastern states. It is better suited to dry situations than the elms or any other maple, and its habit is always erect and compact. Its autumn color is a special feature.

The Norway maple has the same qualities in about equal degree. Its head is lower and broader, its early growth more rapid, but in our climate it is not so durable.

The sycamore maple also grows rapidly, and becomes a large tree of sturdy habit in favorable situations. But it will not thrive without ample moisture in hot weather.

Our red maple is entirely unfit for street planting. The ash-leaved maple does better ; but good street trees of this sort are rare except in the prairie states, where under the name of box elder it is the best and most common street tree. In deep soil its growth while young is remarkably rapid.

In many of our northern towns the European horse-chestnut rivals the elms and the maples. When in full bloom it is gorgeous, but when the fruit is ripe it is especially liable to damage by boys threshing down the nuts. Its foliage is remarkably fine in the early part of the season, but later it is often much preyed upon by caterpillars. In a dry autumn it blisters readily and is sometimes shed prematurely. Usually its shade is too deep for any greensward, and its formal outline is rarely relieved by any grace of form. The Ohio buckeye is not much used compared with the foreign species, probably because its flowers are not so showy, but its average habit is as good and its foliage is better.

Both the European and the American lindens have been extensively planted, especially the former, though inferior in every respect except the fragrance of its flowers. Both do well where the soil is deep and moist and the atmosphere clear. But usually the growth looks stunted and the foliage is more ravaged by caterpillars than any other tree.

The white ash and the European ash are occasionally seen in good condition, but all the other species of ash are fit only for swamps. Their fibrous roots lie too near the surface, and require too much breadth of space.

The beech has similar faults, with the additional defects that its early growth is slow and it requires extra care in transplanting. The European beech is equally disappointing while young, but in good soil its roots do not spread so near to the surface, and good street trees of this species are sometimes seen. But like the purple-leaved and the weeping forms, it is better suited to private grounds.

Walnut, butternut, chestnut and every species of hickory are all handsome trees of good size, but rarely seen in any street except in village roadsides. All require to be transplanted while so young and small that their protection until well established is very difficult. Even if success be finally attained, they will be seriously damaged every fall when they come into bearing, and so their doom is fixed.

Oaks require the same early transplanting and extra care for a long time. But in any city where they can be securely protected until of good size, they endure the ordinary street conditions as well as elms or maples. Once well established a black, red, or scarlet oak will grow as fast as the average of other street trees. The habit is always good, pruning or thinning of the branches is rarely necessary, and the glossy foliage is a special attraction all through the season. The European oaks and our white oak make good street trees if well guarded during their tender years, but their growth is rather slower.

In certain sections of every town the poplars are still the popular trees. All are easily propagated, grow very fast, sell very cheap, endure the most careless transplanting, and thrive in spite of neglect or abuse. But for ordinary streets they are likely to become too large and their life is short. The Lombardy poplar, though a foreign tree and not quite suited to our climate, is the most commonly planted. Its trim, erect habit adapts it to narrow ztreets, and if need be it bears well the most savage lopping of its branches. Where they have ample room to grow to full size, abele or silver-leaved poplar, the balsam and the cottonwood look well, and their smooth, glossy foliage is a special attraction.

But for foliage effect the finest trees are the tulip and the plane. In their native haunts no other deciduous tree equals them in size or dignity. Fine examples of each are occasionally seen in our streets, but general experience seems to condemn them. The soft roots of the tulip tree make it impatient of careless transplanting unless very young, and protection from severe frosts is necessary in clay soils until the roots get below the frost line. But once well established in any favorable soil and subsoil, it becomes a noble street tree, well worth any extra care bestowed upon it.

The plane tree is as easily transplanted as any maple, and if in good soil its growth for many years is as rapid as that of any poplar or willow. But mature trees are so subject to serious fungous blight, that a healthy clean-branched tree is rarely seen. Strangely enough, it thrives well in European cities, withstanding the effects of smoke better than any other tree. In spite of these defects, both tulip and plane deserve persistent trial and experiment.

Fifty years ago, during the silk-worm craze, the Chinese ailanthus was extensively planted in the eastern cities. Its rank growth, sub-tropical aspect, exemption from insects, and its fresh foliage in spite of prolonged heat or drouth, made it very popular. Then came a strange reaction, so strong that the tree is virtually tabooed, all apparently because the flowers have an unpleasant odor. This odor has not a wide range, and it cannot be offensive to many people. Planted in ornamental grounds or near cultivated fields, the tree multiplies so fast, both by seeding and by suckering, that it is often a veritable nuisance. But no tree has withstood so persistently the onslaught of all the destructive influences of a crowded street. Where the subsoil is porous its roots penetrate to an extraordinary depth, and thus find food enough under the closest pavements, and moisture enough during the longest drouths.   There is a place for the ailanthus in every large city, and that (if you give it no other) is the place where no other tree will thrive.

Here and there we occasionally see a frontage or even a whole block planted with trees wholly unsuited to a stony pavement. Among these I venture to class white birch, magnolia, paulownia, yellow locust, honey locust, sweet gum, sour gum, weeping beech or willow, mulberry, cherry and all kinds of pines, spruces, and such like. Some people are charmed with the novelty of the experiment, some are struck by the audacity of the innovation, and some are shocked by the foolishness of the freak.   Comment on each example is as various as the extremes of individual taste.   " Did you see yonder avenue of white birch ?   What a beautiful border for a public street !   Such a happy thought had a touch of inspiration !"   "Ah, yes ! what a fancy frill for a dirty pavement !   How can you bear to see the ' Lady of the Woods ' degraded' to such ignoble use ?   Her white robe all frayed and soiled !   Her ' fragrant hair ' stained with soot and smelling of the gutter !   Away with such defilement of angelic purity!" And so the talk goes on.   You cannot please everybody.   Follow your own bent.

So much has been said on so many kinds of trees that little space is left for comment on their general culture and care.   This must be condensed into a few words on the importance of planting only young trees of nursery growth, of providing ample supply of good soil, of allowing sufficient space for the full growth of root and branch, of watering thoroughly for years, and of guarding from damage at all times.   Americans despise "the day of small things."   This national foible is always prominent in the selection of trees for street for street planting,   The general practice is to procure the largest trees that can be obtained, and conveniently handled.   If nurserymen cannot or will not furnish them of suitable size, they are procured from the neighboring woods if possible.   It is surprising and mortifying to every experienced grower of trees to see each spring the numerous wagons loads which countrymen bring in from swamps and thickets aud expose for sale in our streets day after day, with little or no protection from sun and wind. They are usually much larger than the most overgrown nursery stock, and the younger sap- lings twice or thrice the height becoming to their age, but they are bought in preference to the nurserymen's "small fry." The only roots are a few stout prongs, and they are set out in the smallest holes that will admit them, with the tree tops left unpruned or en- tirely chopped off. They remain standing like bean poles for one or more years! Then they are pulled out and other bean poles stuck in their places. It is said "experience teaches fools," but on this subject they need many years of schooling, else the class always under instruction would not be so large.

Nurserymen preach against this practice incessantly. But the fools think they see through their selfish tricks, and are too wise to be gulled so easily. But they cannot see the potent facts that trees grown in nurseries have needful qualities of root, stem and branch entirely lacking in the spindling saplings that have struggled for life and light in a shady thicket. The nursery plant is forced into vigor- ous growth from the start by providing rich soil, ample space, fre- quent cultivation to induce fibrous roots near the stem, and special training of the plant to an ideal standard of strength and symmetry. When sold it can be dug up with most of its roots uncut, and the small wounds are easily healed. Thus the risk of death by trans- planting is very slight, and with careful work there need not be much check to the tree, but yet the younger the better. Of course after being planted, the smaller the tree the greater the risk of serious dam- age by accidents that would be trifling to one of twice or thrice the size. This argument is the clincher in all discussions on this point. For this reason elms, maples, horse chestnuts, poplars and lindens are so commonly preferred, as they can be successfully transplanted of a much larger size than tulip trees, oaks, or any of the nut bear- ing trees. Yet the rule holds good even in street planting, that what- ever kinds of trees be selected, the youngest that can be protected with a reasonable chance of safety ought to be preferred.

In street planting, sufficient account is rarely taken of the fact that if no grade has been established the position may be too high or too low or too near the future curb line. Or if the street be graded, of

the fact that soil and subsoil may have been removed to such an
extent that the hole has to be dug in hardpan or some substratum
equally sterile.   In such case the quantity of poor ground that must
be dug out and of good soil substituted to secure something like natural conditions is seldom fully appreciated.   Another fact likely to
be overlooked is that an ordinary sidewalk sheds water like a duck's
back, and special provision must be made to supply sufficient moisture during each season of vegetation for years.   At first mulching is
of great advantage to this end.   But in dry weather an occasional
copious watering should be given systematically until the tree is fully
established, and in certain situations this must be continued to some
degree perpetually.   In all such operations the best rule is, never
sprinkle, always soak.   Occasional saturation to the extremities of all
the roots is better than frequent sprinkling.   Another common error
is planting too near the curb line and too close together in the row.
Any young tree within four feet of the curb is ten times more likely
to be gnawed by horses than one twice as far back.   The roots also
should be considered, and given a fair chance to spread on all sides.
Ample distance apart contributes not only to the health and symmetry of the tree, but also allows a pleasant play of sunshine and breeze
to the people on the street.   Close planting may look best for a few
years, but the spread of the trees at maturity should always be pro-
vided for.   The future cutting out of each alternate tree is a lovely
illusion fondly embraced, but never realized; but in reality a sad
delusion because it is rarely done, and never done soon enough.   It
may be a deeply inspired full-blown resolve, but it has no substance.
It is only a soap-bubble that floats before you for a moment, and in-
stantly melts out of sight.   Some protective guard against ill-bred
horses, worse-bred drivers, careless workmen on the street or adjacent lots, and the daily run of miscellaneous accidents, is necessary
for years.   Nothing yet invented is conveniently applicable to small
trees or always effective.   A temporary railing on the curb line, though
unsightly, is more useful than a casing for each tree.   When the
trunk becomes thick enough, a strip of fine galvanized wire netting
wrapped loosely around it as far up as a horse can reach is cheap,
serviceable, neat, unobtrusive, and can readily be adjusted to the
growth of the tree from year to year.   The damage done to street

trees by horses and by careless usage of workmen about them is incalculable. Prosecution is useless, because an adequate penalty that would deter others is never imposed.

It is needless to speak to experienced fruit-growers of the best preventives or remedies for insect pests of any kind, or for any of the various kinds of fungus. That work is as familiar to you as the alphabet. But systematic and effective treatment of street trees for any plague that may attack them is exceedingly rare and difficult. Some caterpillars are easily conquered if the eggs or the chrysalis be attacked in time, but this is usually prevented by stoical indifference at this dormant stage. Disease caused by smoke and dust is incurable. No species of tree is proof against these if their presence be nearly constant in large quantities. Finally, the gist of the whole matter may be summed up in the obtrusive form of sententious advice. Select the kinds of trees that experience commends to you as most likely to satisfy your own taste, and to suit the conditions of soil and situation in your street. Select young trees only, of thrifty habit and good form. Furnish good soil in ample quantity at whatever cost or trouble. Give ample room for full growth on every side. Handle and transplant with proper care and skill. Mulch and water effectively until the trees be fully established. Guard from damage by any device that will serve your purpose. Fight to the death every pest and plague as soon as it appears. Give constant watchfulness to the tree's welfare while you live, and impose the same duty by your last testament upon the successors to your trust when you die. '' Eternal vigilance is the price '' of every street tree.

A Good Book.

Be careful to write name and post office plainly, so that there may be no mistake in mailing.   Address

THE RURAL PUBLISHING CO., *New York.*

THE NURSERY BOOK.—By L. H. BAILEY.   A complete handbook of Propagation and Pollination of Plants. *Profusely illustrated.* This valuable little manual has been compiled with great pains. The author has had unusual facilities for its preparation, having been aided by many experts. The book is absolutely devoid of theory and speculation. It has nothing to do with plant physiology or abstruse reasoning about plant growth. It simply tells, plainly and briefly, what every one who sows a seed, makes a cutting, sets a graft, or crosses a flower wants to know. It is entirely new and original in method and matter. The cuts number 107, and are made expressly for it, direct from nature. The book treats of all kinds of cultivated plants, fruits, vegetables, greenhouse plants, hardy herbs, ornamental trees, shrubs and forest trees.

CONTENTS:

I.—SEEDAGE. On Propagation by Seed.
II.—SEPARATION.
III.—LAVERAGE. Propagation by Layering.
IV.—CUTTAGE. Propagation by Cuttings.
V.—GRAFTAGE.—Including Grafting, Budding, Inarching, etc.
VI.—NURSERY LIST.—This is the great feature of the book. It is an alphabetical list of all kinds of plants, with a short statement telling which of the operations described in the first five chapters are employed in propagating them. *Over 2,000 entries* are made in the list. The following entries will give an idea of the method:

Acer (MAPLE). *Sapindaceæ.* Stocks are grown from stratified seeds, which should be sown an inch or two deep; or some species, as *A. dasycarpum*, come readily if seeds are sown as soon as ripe. Some cultural varieties are layered, but better plants are obtained by grafting. Varieties of native species are worked upon common or native stocks. The Japanese sorts are winter-worked upon imported *A. polymorphum* stocks, either by whip or veneer grafting. Maples can also be budded in summer, and they grow readily from cuttings of both ripe and soft wood.

Phyllocactus, Phyllocereus, Disocactus (LEAF CACTUS). *Cacteæ.* Fresh seeds grow readily. Sow in rather sandy soil which is well drained, and apply water as for common seeds. When the seedlings appear, remove to a light position. Cuttings from mature shoots, three to six inches in length, root readily in sharp sand. · Give a temperature of about 60°, and apply only sufficient water to keep from flagging. If the cuttings are very juicy they may be laid on dry sand for several days before planting.

VII.—POLLINATION.

Price, in Library Style, cloth, wide margins, $1. Pocket Style, paper, narrow margins, 50 cents.

The Rural Library.

✦ ✦ ✦

HE Rural Library is a series of monthly issues of popular pamphlets on scientific and practical topics in agriculture and horticulture. Each is intended to be a complete manual of the subject treated. They are written or edited by men and women whose broad knowledge of the specialties on which they write is undisputed. The range of subjects is well illustrated by the titles of the numbers already issued or soon to appear. Among these are: "Strawberries: What, How and Wherefore;" "Tools: Styles, Types and Methods for use in Horticulture;" "Cross-breeding and Hybridizing of Plants;" Chemicals and Clover;" "The Business Hen;" "The Tuberous Begonia;" "Rats and Other Vermin;" "Canning and Preserving Fruits and Vegetables;" "The New Celery Culture;" "Ensilage and the Silo;" "Accidents and Emergencies," and "Milk: Making and Marketing." Others to come will, in turn, enter upon all departments of rural activity wherein more extended information is required than the newspaper article can furnish, or wherein existing books fail to condense properly for handy use the particular information wanted. •

THE RURAL PUBLISHING COMPANY,

Times Building, New York.